the SCIENCE *library*

HOW THINGS WORK

the SCIENCE *library*

HOW THINGS WORK

John Farndon
Consultant: Steve Parker

This 2009 edition published and distributed by:

Mason Crest Publishers Inc.

370 Reed Road, Broomall, Pennsylvania 19008

(866) MCP-BOOK (toll free)

www.masoncrest.com

Library of Congress Cataloging-in-Publication data is available

How Things Work
ISBN 978-1-4222-1550-0

The Science Library - 10 Title Series
ISBN 978-1-4222-1546-3

Printed in the United States of America
First published in 2004 by Miles Kelly
Publishing Ltd
Bardfield Centre Great Bardfield
Essex CM7 4SL
Copyright © 2004 Miles Kelly Publishing Ltd

Editorial Director Belinda Gallagher

Art Director Jo Brewer

Editor Jenni Rainford

Editorial Assistant Chloe Schroeter

Cover Design Simon Lee

Design Concept Debbie Meekcoms

Design Stonecastle Graphics

Consultant Steve Parker

Indexer Hilary Bird

Reprographics Stephan Davis, Ian Paulyn

Production Manager Elizabeth Brunwin

Contents

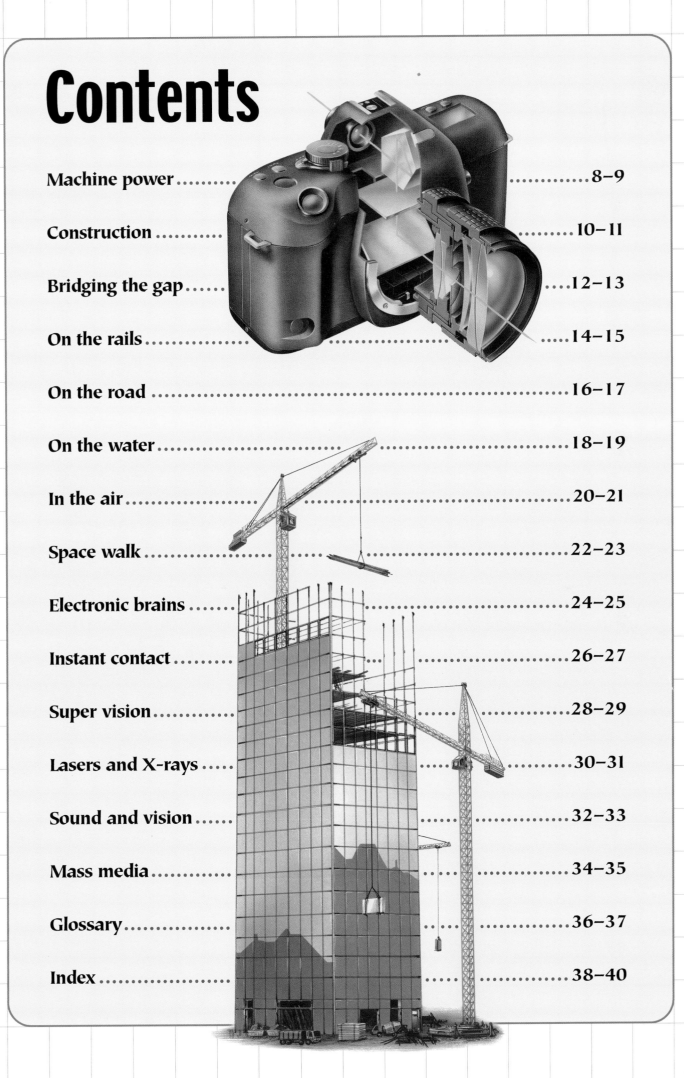

How to use this book

HOW THINGS WORK is packed with information, color photos, diagrams, illustrations and features to help you learn more about science. Do you know how your camera takes pictures or how skyscrapers are built? Do you know what is inside a computer or that light from lasers is brighter than light from the Sun? Enter the fascinating world of science and learn about why things happen, where things come from and how things work. Find out how to use this book and start your journey of scientific discovery.

It's a fact
Key statistics and extra facts on each subject provide additional information.

Main text
Each page begins with an introduction to the different subject areas.

Cross-references
Attached to captions and pictures are cross-references that use the unique co-ordinates grid system. These lead you to related subjects within the book.

Main image
Each topic is clearly illustrated. Some images are labeled, providing further information.

Check it out!
Find out more by surfing the Internet.

10

Construction

IN THE past, most big buildings such as cathedrals were built of stone, and their strength lay in the thick walls. Today, buildings such as skyscrapers get their strength from a framework or 'skeleton' of steel girders, beams and concrete columns, forced into the ground by heavy machinery such as pile-drivers. The skeleton supports the roof, walls and floors so the walls can even be made of glass, if designed to carry no weight. Tunnels and arches are different types of construction with their strength in curved walls or arches.

IT'S A FACT
- The world's longest tunnel carries water for 169 km, from a reservoir direct into New York City.
- The world's longest underwater tunnel is Japan's Seikan tunnel, which carries a railway 53.9 km under the Tsugaru Strait between Honshu and Hokkaido islands.

To scale
Two of the world's tallest struc
1 square = 100 m high

452 m high

Petronas Twin Towers
(Kuala Lumpur, Malaysia)

5. Cladding of glass and concrete walls are added

4. Lifts and other services such as electricity and plumbing are installed

3. Concrete and steel frame floors or decks are slowly added, starting from the bottom

2. Heavy steel girders are lifted into position by tall cranes as the building rises

1. Foundations are dug and piles are set in the ground

Scraping the sky
Skyscrapers are dramatic buildings, often towering many hundreds of metres into the air. To support such a huge structure, every skyscraper must have deep foundations (the part below ground) made of massive concrete or steel piles sunk into the ground. Above these, the walls and floors are attached to a skeleton of steel girders and beams. When a skyscraper is over 40 storeys (levels) tall, the weight of a strong wind blowing against the side of the building is greater than the weight of the actual building. So architects must make the building strong horizontally as well as vertically.

▶▶ Read further › steel girders
pg13 (k22)

Check
- http://ww
buildingl
- http://w
com/

The world's tallest structure is the 553-m CN tower in Toronto, Canada

1 2 3 4 5 6 7 8 9 10 11 12 13 14 15 16

Photos and artworks
Illustrations and photographs accompany each caption. Diagrams are labeled to give more detailed scientific facts and information.

The grid
The pages have a background grid. Pictures and captions sit on the grid and have unique co-ordinates. By using the grid references, you can move from page to page and find out more about related topics.

Amazing facts
Look out for facts that run along the bottom of each page.

To scale
This feature uses the grid to show comparative size of different objects. You can easily compare and see exactly how small or large things are.

WORK IT OUT

• The first skyscrapers were built in New York and Chicago, USA. The 10-storey Home Insurance building in Chicago, was completed in 1885.

• The invention of the safety elevator by Elisha Otis in the 1850s made skyscrapers possible. Today, elevators in skyscrapers can travel at speeds of up to 10 m/sec.

-drivers
ngs or bridges erected on soft ground are
ed by posts called piles. Usually made of steel
rete, the piles are driven firmly into the ground
-drivers. Pile-drivers wind up a heavy weight
a pile hammer inside a frame, then drop
he head of the pile to bash it into the
d. In 1847, Scottish engineer James
mith invented a steam-driven
on that could raise and lower
onne weight 80 times a
ute. Today, pile-drivers
compressed air to
se the hammer, and
e drop is precisely
ontrolled by computers.

Read further > compressed air
pg19 (b29); pg21 (b31)

Going underground
Tunnels just below the surface can be built by the 'cut-and-cover' method. This means simply digging a large, long ditch, then covering it over to make the tunnel. But deeper tunnels must be bored out. Tunnels through hard rock are often blasted out using explosives. Tunnels through soft rock or soil are dug with a powerful cutting machine called a shield, which is a large drum with disk cutters on the front. As the shield bores forward, debris is scooped out the back. Ring-shaped steel and concrete supports are put in place to prevent the tunnel caving in.

Keystone block in upper centre of arch 'locks' structure in place by balancing pressure from each side

Building an arch
Arches are usually made of stone or brick and built up from wedge-shaped blocks called voussoirs. While the arch is being built, a wooden frame the same shape as the arch supports the structure. When the frame is removed, the arch is supported by the pressure of each side of the arch against the keystone (centre block).

Brick or stonework sides prevent the arch collapsing under the weight of structure above

◀ The world's biggest free-standing arch is the 192 m-high Gateway to the West arch in St Louis, Missouri, USA.

▶ The shield's rotating cutting head is kept on a straight course by lasers.

Read further > bridges / lasers
pg12 (d2); pg30 (d15)

Read further > arch bridges
pg12 (s8)

In 1931 New York's Empire State Building was completed in a record time of 15 months

20 21 22 23 24 25 26 27 28 29 30 31 32 33 34 35 36 37 38 39

a b c d e f g h i j k l m n o p q r s t u v w

Machine power

MACHINES MAKE it easier to perform tasks that would otherwise be very difficult or even impossible. Machines carry out tasks more efficiently by changing the strength or direction of a force. A machine can be as simple as a door handle or as complex as a spacecraft (*see pg22 [p5]*) but all include types of six basic machines: the ramp or inclined plane, the wedge, the lever, the screw, the pulley, the wheel and axle. But most machines, even the complex ones, work because of the link between force and distance.

(see pg22 [p5])

IT'S A FACT

• Humans first used levers about 2 million years ago, when handles were attached to axes to gain power for cutting.

• The world's biggest crane is the Japanese floating Musashi that can lift over 3000 tons.

Effort

Load

First-class lever

Effort

▲ *Prizing the lid from a can of paint: the tool acts as a first-class lever with the wrist doing the effort.*

Fulcrum

Gearing up

'Driver' cog is larger with more 'teeth'

Gears are used in machines such as clocks and cars, which have a rotating mechanism (parts that go round). Gears can alter the power from an engine so that it moves at the right speed and force, and goes in the right direction. Gears are usually sets of wheels called cogs with teeth that 'mesh' and turn together (*see pg16 [u8]*). A big cog turns a smaller cog with less force but faster. A small cog turns a bigger cog slower but with more force.

'Driven' cog has fewer teeth and turns slower but more powerfully

Read further > cars / gears
pg16 (b15; p8); pg17 (l22)

How levers work

A lever is a bar or rod that is supported by a fulcrum (central pivot). Effort or force is needed to move one end of the lever to move a load at the other end. The closer the fulcrum is to the load, the less effort is needed to lift it. There are three classes of lever. In first-class levers, such as a seesaw, the fulcrum is in the middle. Second-class levers, such as a wheelbarrow have the fulcrum at one end, the effort (pushing) at the other end and the load in the middle. In third-class levers, such as a hammer, the effort is between the load and the fulcrum (the hand).

Wheel power

A heavy load is very hard to drag uphill on its own but rolling it in a wheelbarrow is much easier. The wheel reduces the friction (resistance) so that the load can be moved more easily. A ramp makes a task easier by enabling the load to be pulled or pushed up a slope rather than lifted.

Wheel and axle

Load

Effort

Fulcrum

Ramp or inclined plane

Read further > wheels
pg17 (b22)

▲ *Three simple machines: a wheelbarrow (lever), wheel and axle, and ramp (inclined plane).*

The Egyptian pyramids were built by humans using simple machines such as wheels, ramps and pulleys

Weight lifters

A pulley is the best machine for lifting heavy loads. It can be as simple as a rope or chain flung over a bar. Usually, though, the rope runs over a number of grooved wheels. A single pulley wheel helps by changing the direction of the force because pulling down is easier than lifting up. If the rope is looped round and back over several pulley wheels, the load is spread over a great length of rope, and is thus reduced, making the object easier to lift. Pulley 'blocks' have many sets of pulleys and can lift very heavy loads for quite a small effort.

▼ *A simple set of pulleys allows a person to lift a heavy weight using distance to increase the force.*

Effort

More pulleys give a greater force for the person pulling

Load

▶▶ **Read further › weights** pg11 (b22)

Moving staircase

▶▶ **Read further › elevators** pg11 (b32)

Escalators – moving staircases – carry people up or down using the power of a pulley. The escalator's pulley is like a bicycle chain wrapped round a cog-like pulley at either end of the staircase. The stairs go round in a loop, moving down on top and coming back up from underneath, in reverse. The weight of the descending stairs pulls the ascending stairs upwards. So the escalator's drive motors only have to lift the weight of the passengers, not the weight of the stairs, too.

Ascending or descending stairs

Motor-driven pulley

Stairs are fixed to the pulley with brackets

▼ *This seesaw is a first-class lever, as the fulcrum is between the load and effort.*

Load

◀ *Corkscrews are a type of turning wedge.*

▲ *An escalator's steps swivel past each other at either end of the pulley system.*

▶▶ **Read further › weight of loads** pg12 (d2)

▶▶ **Read further › propellers** pg18 (i13)

● WORK IT OUT

• Many bones in the human body work like levers. For example, the forearm pivots at the elbow, acting as a type of third-class lever to give extra lifting or pushing power.

• Bicycle gears help cyclists pedal between 60 to 90 turns a minute, reducing the effort needed.

Simple machines

Two ramps (inclined planes) back to back form a wedge, for separating or splitting items, as in an axe. A wedge wrapped around a pole or rod forms a screw, which changes a turning force into a slow, steady, pulling force.

Check it out!

• http://www.howstuffworks. com/pulley.htm
• http://www.fi.edu/qa97/ spotlight3/spotlight3.html
• http://www.galaxy.net:80/~k12 /machines/index.shtml

A 213 m-tall crane was used to help build the First National Centre Tower at Omaha, Nebraska, USA

Construction

IN THE past, most big buildings such as cathedrals were built of stone, and their strength lay in the thick walls. Today, buildings such as skyscrapers get their strength from a framework or 'skeleton' of steel girders, beams and concrete columns, forced into the ground by heavy machinery such as pile-drivers. The skeleton supports the roof, walls and floors so the walls can even be made of glass, if designed to carry no weight. Tunnels and arches are different types of construction with their strength in curved walls or arches.

IT'S A FACT

- The world's longest tunnel carries water for 169 km, from a reservoir direct into New York City.

- The world's longest underwater tunnel is Japan's Seikan tunnel, which carries a railway 53.9 km under the Tsugaru Strait between Honshu and Hokkaido islands.

Scraping the sky

Skyscrapers are dramatic buildings, often towering many hundreds of meters into the air. To support such a huge structure, every skyscraper must have deep foundations (the part below ground) made of massive concrete or steel piles sunk into the ground. Above these, the walls and floors are attached to a skeleton of steel girders and beams. When a skyscraper is over 40 storys (levels) tall, the weight of a strong wind blowing against the side of the building is greater than the weight of the actual building. So architects must make the building strong horizontally as well as vertically.

▶▶ Read further > steel girders
pg13 (k22)

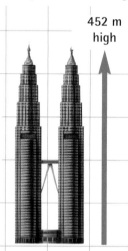

5. Cladding of glass and concrete walls are added

4. Lifts and other services such as electricity and plumbing are installed

3. Concrete and steel frame floors or decks are slowly added, starting from the bottom

2. Heavy steel girders are lifted into position by tall cranes as the building rises

1. Foundations are dug and piles are set in the ground

To scale
Two of the world's tallest structures
1 square = 100 m high

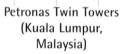

452 m high

300 m high

Petronas Twin Towers (Kuala Lumpur, Malaysia)

Eiffel Tower (Paris, France)

Check it out!

- http://www.pbs.org/wgbh/ buildingbig/skyscraper/
- http://www.skyscraperpage. com/

Pile-drivers

Buildings or bridges erected on soft ground are supported by posts called piles. Usually made of steel or concrete, the piles are driven firmly into the ground by pile-drivers. Pile-drivers wind up a heavy weight called a pile hammer inside a frame, then drop it on the head of the pile to bash it into the ground. In 1847, Scottish engineer James Naysmith invented a steam-driven version that could raise and lower a 7-ton weight 80 times a minute. Today, pile-drivers use compressed air to raise the hammer, and the drop is precisely controlled by computers.

Read further › compressed air
pg19 (b29); pg21 (b31)

WORK IT OUT

• The first skyscrapers were built in New York and Chicago, USA. The 10-story Home Insurance building in Chicago, was completed in 1885.

• The invention of the safety elevator by Elisha Otis in the 1850s made skyscrapers possible. Today, elevators in skyscrapers can travel at speeds of up to 10 m/sec.

Going underground

Tunnels just below the surface can be built by the 'cut-and-cover' method. This means simply digging a large, long ditch, then covering it over to make the tunnel. But deeper tunnels must be bored out. Tunnels through hard rock are often blasted out using explosives. Tunnels through soft rock or soil are dug with a powerful cutting machine called a shield, which is a large drum with disk cutters on the front. As the shield bores forward, debris is scooped out the back. Ring-shaped steel and concrete supports are put in place to prevent the tunnel caving in.

Building an arch

Arches are usually made of stone or brick and built up from wedge-shaped blocks called voussoirs. While the arch is being built, a wooden frame the same shape as the arch supports the structure. When the frame is removed, the arch is supported by the pressure of each side of the arch against the keystone (center block).

Keystone block in upper centre of arch 'locks' structure in place by balancing pressure from each side

Brick or stonework sides prevent the arch collapsing under the weight of structure above

◀ *The world's biggest free-standing arch is the 192 m-high Gateway to the West arch in St Louis, Missouri, USA.*

▶ *The shield's rotating cutting head is kept on a straight course by lasers.*

Read further › arch bridges
pg12 (s8)

Read further › bridges / lasers
pg12 (d2); pg30 (d15)

In 1931 New York's Empire State Building was completed in a record time of 15 months

a b c d e f g h i j k l m n o p q r s t u v w

Bridging the gap

Central tower has two cantilever 'arms'

Steel beams span length of bridge

Piers on piles support weight of main towers

BRIDGES ARE vital links, carrying traffic – people, animals and vehicles – over rivers, roads, railways or ravines. To be safe, a bridge must be strong enough to bear the weight not only of the load moving across it, but also its own weight. This strength can be achieved in a number of ways, but every bridge has massive supports at either end called abutments that are embedded in solid ground. In between there are often extra supports called piers, linked by short sections of the bridge called spans. There is usually a long, middle span over the deepest water.

LONGEST SUSPENSION BRIDGES

Bridge, Country	Length
Akashi–Kaiko, Japan	1990 m
Great Belt, Denmark	1624 m
Humber Estuary, Britain	1410 m
Jiangyin, China	1385 m
Tsing Ma, China	1377 m

Types of bridge

Traditionally, bridges were simple wooden or stone arches. Modern bridges are built of concrete and steel, and the type of construction depends on the heaviest load it has to carry and the nature of the place where it is to be built. For example, for wide spans over water, suspension or cable-stayed bridges are usually best. To carry a very heavy load over a short distance, a beam bridge may be built. To carry a heavy load over a greater distance, a cantilever bridge is often used.

Read further › construction pg10 (k2)

▼ Cable-stayed bridges, such as the Queen Elizabeth II Bridge, Dartford, hang from steel cables.

▶ Suspension bridges, such as the Golden Gate Bridge, San Francisco, have spans that hang on steel wires from a cable between two tall towers.

▼ Lifting bridges, such as Tower Bridge, London, use hydraulic power to swing a section of the bridge up to allow tall ships to pass underneath.

▶ Arch bridges, such as London Bridge, are very strong because the weight is pushed outwards to the abutments, rather than downwards.

▲ Cantilever bridges, such as the Forth Railway Bridge, Scotland, balance the load on two piers.

▲ Beam bridges, such as most city road bridges, rely on the strength of the material – usually steel or concrete – to support a beam between the spans, strengthened by a framework underneath.

Check it out!

- http://www.pbs.org/wgbh/ buildingbig/bridge/challenge/ indexp.html
- http://www.howstuffworks. com/ bridge.htm

The Anji bridge at Zhao Xian in China is 1400 years old

Cantilever bridges

Cantilever bridges are built from sections, each with two rigid steel beams on either side of a central tower. Each pair of beams balance each other so that each section of the bridge is an independent cantilever, standing without support from the others. The towers support the weight of the beams with a downward force, so there is no need for central supporting columns.

Supporting tower carries weight of link to bank

Read further > beams
pg10 (d2)

▸ *In a bascule or lifting bridge, the half-spans are supported only at one end. The middle tilts up out of the way.*

Swinging around

If the banks of a river or canal are very low, one way to ensure that boats can pass underneath without colliding with the bridge is to move the actual bridge. Bascule bridges are built to lift up in the middle. Swing bridges are mounted on a central pier and pivot sideways out of the way, turned by powerful hydraulic (fluid-driven) systems.

Read further > boats
pg18 (d2; i13)

Read further > pumping water
pg18 (o16); pg19 (b25)

Building bridges

The first step in bridge-building is to build the piers and abutments. To erect piers in the water, a steel wall or shuttering is built on the riverbed. The water is then pumped out while the pier is built, before the walls can be removed, and the spans laid between the piers.

Steel wall (shuttering)

Water

Collar

Space to accept pier or tower

Footings

River bed

IT'S A FACT

• The largest tilting or lifting bridge is in Michigan, USA. It has a 102-m opening.

• The world's longest arch is the 550 m-long Lupu steel arch in Shanghai, China.

Support tower

Main suspension cable

Vertical stay cable

Suspension

For wide or deep channels, a suspension bridge, with a span of up to 1200 m, is often used. The deck (such as a roadway) is held up by steel cables hanging between tall towers. These cables are held in concrete blocks at both ends of the bridge. The deck may be supported by a framework to stop it swaying too much in the wind. The weight of the deck is transferred to the vertical cables, over the towers and to the concrete blocks at either end.

Read further > frameworks
pg10 (d2; k2)

Support pier

Deck

Nine of the world's 20 longest bridges are in Japan

a b c d e f g h i j k l m n o p q r s t u v w

On the rails

A SINGLE TRAIN may carry 200,000 tons of iron ore, while many passenger trains have many carriages that can carry thousands of people. Railways can carry much heavier loads than road vehicles because they use steel wheels, guided by lips or 'flanges' that run along the inside of steel rails, which are tough enough to bear tremendous weight. However, not all rail transport runs on the rails – monorails often run underneath the track and maglevs actually float above it. Because they travel at great speeds, trains must be safely controlled by switches and signals.

● WORK IT OUT

• The world's fastest scheduled service is the Shinkansen bullet train from Hiroshima to Kokura in Japan. It covers the 192-km distance in just 44 minutes, averaging 261.8 km/h.

• Across the world, there is about 1.3 million km of railway track.

● One-track trains

Monorails are railways with a single rail. The track is made from steel or concrete sections and the carriages either hang beneath the rail or balance on top of it. When the cars run on top of the rail, they must be stabilized by guide wheels and gyroscopes – devices that rotate – to stop them falling over. Monorails are over a century old – the first was built in Wuppertal in Germany in 1901. Both Tokyo and Seattle have monorail systems in operation.

▶▶ **Read further › underground tunnels pg11 (g33)**

▶ *Monorail tracks can be held up above city streets, avoiding the need to dig deep tunnels.*

● IT'S A FACT

• The longest train ever was a 660-truck goods train 7.3 km long that ran from Saldanha to Sishen in South Africa on 26 August 1989.

• The longest passenger train ever was a 70-coach train 1732 m-long that ran from Ghent to Ostend in Belgium on 27 April 1991.

● **Check it out!**

• http://travel.howstuffworks.com/maglev-train.htm
• http://www.csrmf.org/

Gyroscope sensors

Stabilizing side wheels

Monorail section

Weight-bearing drive wheels made of rubber for less noise and vibration

● Signalling

Signals warn train crews of hazards on the track. Track controllers use a 'block' system to stop a train entering a block of track that already has a train on it. Some routes in Europe and Japan use Advanced Train Protection (ATP), where the cab picks up signals from the track telling the driver what speed to travel. If the driver fails to respond, the train slows automatically. In the US, they are developing Advanced Train Control Systems (ACTS) that will rely on satellites and other high-tech links.

▲ *Signal lights are usually red for stop or danger, yellow for warning and green for all clear.*

▶▶ **Read further › satellites**
pg27 [k22]; pg27 [p26]

● Lifting by magnets

When the same poles of two magnets are placed together, they repel (push each other apart). Magnetic levitation (maglev) trains use this magnetic repulsion to lift the train clear of the track. Very powerful electromagnets lift the entire train so that it floats a few centimetres above the track. This cuts the friction so much that maglev trains can reach speeds of more than 480 km/h smoothly and almost silently. Plans are underway to build fast maglev lines from Anaheim in California to Las Vegas, and from Tokyo to Osaka in Japan, but they are costly to build and still not totally reliable. So most maglev systems have been short distance, low-speed train systems, such as the system at Shanghai Airport in China.

▶▶ **Read further › magnets**
pg19 [h22]; pg31 [b32]

Track electromagnets

Concrete T-section track

▲ *Maglev trains operate in many countries such as China, Japan, Germany and the United States.*

Support pillar

Train electromagnets

Old track direction

Frog to carry train across old track

Switch rail pivots

Guard rail

New track direction

Switch rail pivots

◀ *Switches are used to automatically change the direction of a railway train.*

Slider moves switch rails across

Switch rails

● Points and junctions

Occasionally, a train must change tracks in order to change direction. As a train cannot steer, the track branches in two at 'turnouts.' Here a switch changes the train from one track to another by moving a pair of rails for the new track direction up against the old. The two switch rails pivot at one end and are moved sideways by a slider under them at the other end. The slider is usually moved by an electromagnet.

▶▶ **Read further › steering wheels**
pg17 [v30]

The gap between the rails, called the gauge, is widest in Spain, Portugal and India, at 1.67 m

On the road

THERE ARE now more than half a billion cars in the world, and on average two new ones are built every second. There are a huge variety of motor vehicles, from cars, to 'all-terrain' vehicles that can be driven cross-country, and motorcycles that can reach great speeds. However, all motor vehicles are built in a similar way. They all need brakes, are all driven by an engine or motor and most need gears to control the 'drive' of the engine.

Brake caliper

Brake fluid under pressure in brake pipe pushes on piston

Brake piston

Piston presses two brake pads together

Putting on the brakes

Stopping a fast-moving car demands great force. When the driver presses the brake pedal, brake fluid is forced through narrow pipes to cylinders on each wheel. The pressure of the fluid presses special pads – brake pads – against the brake discs on the wheels, which slow it down by friction. Many cars have ABS (Anti-lock braking systems) where a computer applies the brake automatically within a split second and prevents the car's wheels from locking up and the car from skidding to a halt.

Brake pads press on brake disc to create friction, thus slowing down wheel

Read further › brake pads
pg17 (v26)

Read further › gears
pg8 (n2)

Geared up

Most cars need gears between the engine and the wheels because engines only work well when running faster within a narrow range of speed. Gears change the drive – the link between engine and wheels – so the wheels turn at different speeds for the same engine speed. Slowing the drive increases the engine's force, enabling a small, fast-running engine to power a heavy car. When accelerating or climbing hills, the driver selects the low gears for extra force. In cars with automatic transmission, the right gear is automatically selected.

WORK IT OUT

• The world's longest car was specially built for American, Jay Ohrberg. At over 30 m long, it has 26 wheels – and a swimming pool with a diving board!

• In 1997, the British jet-powered Thrust SSC car broke the sound barrier, travelling across the Nevada desert at over 1220 km/h.

▲ Simple cogs reverse the direction of rotation.

▲ Rack and pinion change the rotation to a sliding motion.

▲ Bevels move the direction of rotation through a right angle.

▲ Worm gears slow the rotation down to a slower, stronger rotation at right angles.

Over 22 million Volkswagen 'Beetles' have been built since they were introduced in 1937

Leaning into it

To go round corners, motorbikes' front wheels turn like cars do. Motorcyclists also lean the bike over on to the slanting edges of its tires. If a motorcyclist tried to turn a corner with the bike in an upright position the force called momentum, which makes the bike carry on in a straight line, would tip it over onto the outside of the curve. So the rider leans the motorbike when turning the corner to counteract the outwards force pushing against the bike.

◀ *The faster the bike corners, the greater the outward forces become and the more the biker has to lean over to counter them.*

▶▶ **Read further > wheels** pg8 (s8)

Gasoline power

Gas engines burn a mixture of fuel (gas) and air inside cylinders in the engine. The mixture expands (increases in volume) when it is ignited (set alight) by an electric spark, from the spark plug. This forces a piston down the cylinder to push the crankshaft round to turn the car's wheels through the gearbox. Four cylinders (two- or three-cylinder engines exist) give these power strokes, each at a different time.

▶ *Four-stroke cycle: commonly known as 'suck, squeeze, bang and blow.'*

▶▶ **Read further > cylinders / gears** pg16 (k2; p8)

Valves
Cylinder
Piston
Crankshaft

▲ *Induction stroke – inlet valve opens and the descending piston sucks in the fuel and air mixture.*

▲ *Compression stroke – inlet valve closes and the rising piston squeezes the fuel.*

Spark plug

▲ *Power stroke – spark ignites fuel and expanding gases push down the piston.*

▲ *Exhaust stroke – the exhaust valve opens and the rising piston pushes or blows out the burned gases.*

Electrical system powers lights and instruments, such as temperature gauge and fuel gauge, on instrument panel

Exhaust at back of car carries away waste gases from engine

Gears in gearbox change the speed engine turns the wheels to suit different conditions

Ignition system gives an electrical spark to ignite the fuel

Engine burns fuel to provide power

Car transmission takes power from engine to wheels via the gearbox

Cooling system circulates water to stop engine overheating

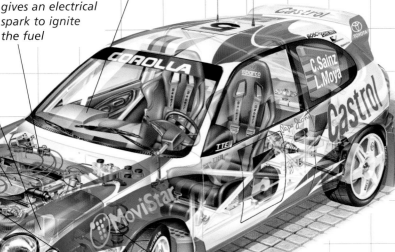

Suspension softens bumps in road for passengers and keeps wheels in contact with road

Steering gear controls direction by turning front wheels

Wheels are attached to brakes, which slow them down

There are enough cars in the world to create a traffic jam stretching right around the world 10 times!

On the water

LIGHT MATERIALS such as wood and cork float because they are not as dense (heavy) as water. Ships can be built from heavy materials such as steel, yet still float. This is because the air inside the hull or body of the ship makes the ship lighter than the water it displaces (pushes out of the way). Water transportion relies on many methods to move through water. Sailing ships use the power of the wind to travel along. Hydrofoils use 'wings' to travel on top of the water. Submersibles and submarines actually let in water to sink, before travelling underwater.

WORK IT OUT

• Speeds at sea are measured in knots, because sailors used to measure their speed by counting the knots passing in a knotted rope let out behind the boat.

• One knot at sea is the equivalent of 1.852 km/h on land.

Metal blades or screws slice through water, propelling boat forward

Rudder steers boat left or right

Thrust and drag

Most ships are driven under the water by a screw propeller with slowly turning blades to force (thrust) the ship forward. This system is powerful, but the resistance of the water slows down the boat (drag). Small powerboats use high-speed jets of water instead to push themselves along faster.

▶▶ Read further › propeller blades pg20 (j12)

IT'S A FACT

• The heaviest ship ever built is the oil tanker, *SeaWise Giant*, launched in 1976, weighing over 585,015 tons when laden.

• The world water speed record is 511.11 km/h set by Kenneth Warby's *Spirit of Australia* on 8 October 1978, using a 'hydroplane' designed to skim across the water.

How ships float

When launched, a ship pushes water away, but the water pushes back, causing 'upthrust.' The more water is pushed out of the way, the harder it pushes back. The hull of the ship is hollow, making the density – weight – less than water, so the ship keeps sinking until its weight is matched by the push of the water, then it can float.

Check it out!

• http://oceanexplorer.noaa.gov/technology/subs/subs.html

Drag boats are small but extremely powerful motorboats that can travel faster than 350 km/h

| 1 | 2 | 3 | 4 | 5 | 6 | 7 | 8 | 9 | 10 | 11 | 12 | 13 | 14 | 15 | 16 | 17 | 18 | 19 |

▼ *The first range of submersibles appeared in the 1960s and 1970s. Today's craft are much smaller and more advanced but still work in the same way.*

● Underwater explorer

Submersibles are used mainly for research into the ocean depths and undersea wrecks. Remote Operated Vehicles (ROVs) are small robot submersibles, controlled by operators using cameras and virtual reality systems *(see pg24 [o12])*. Submersibles alter their buoyancy – ability to float – to work.

Read further › magnets / virtual reality
pg15 [b30]; pg24 [m2]

Propeller for pushing craft through water

Cabin of strong steel to resist intense water pressure

Double hatch containing airlock for divers to exit and re-enter

Powerful electric motor

Searchlights for seeing dark ocean depths

To go back up to the surface, pilot switches off electromagnets that hold ballast of iron balls in place

Claw for grabbing samples during missions

Extra-strong Perspex dome

Float filled with gasoline. Since gasoline is lighter than water, it helps keep craft afloat

Batteries

Small propellers called thrusters maneuver craft precisely up and down and sideways

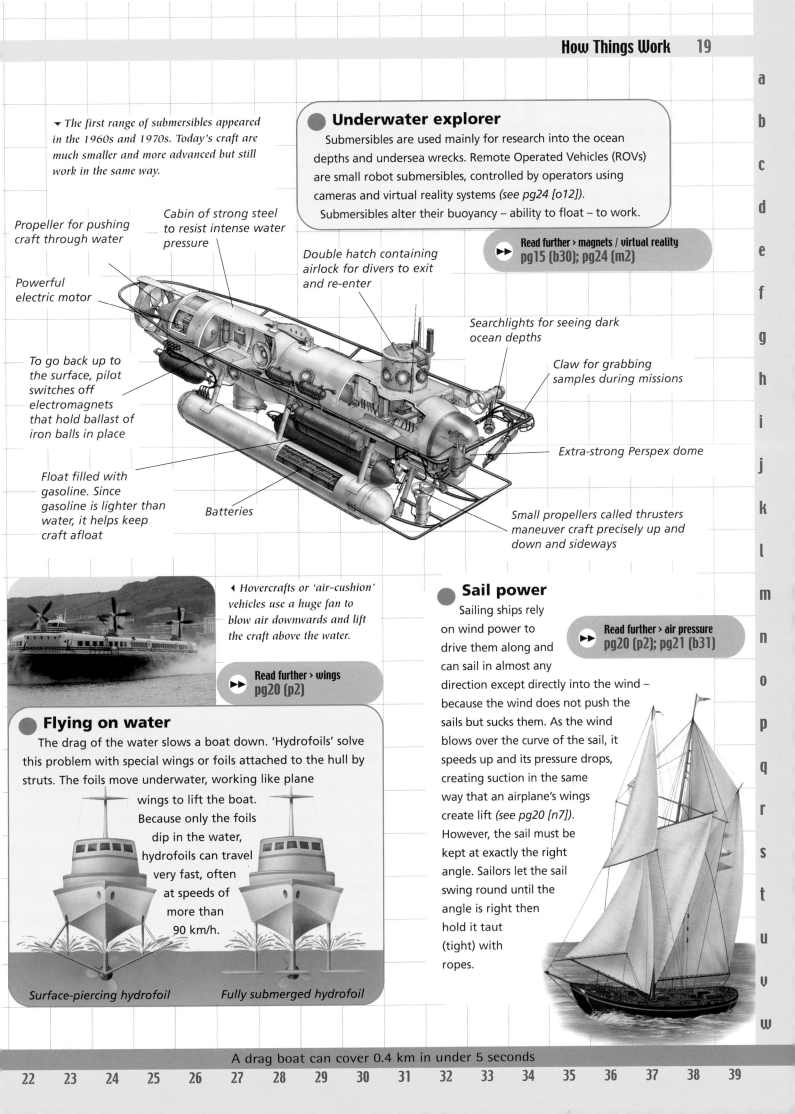

◄ *Hovercrafts or 'air-cushion' vehicles use a huge fan to blow air downwards and lift the craft above the water.*

Read further › wings
pg20 [p2]

● Flying on water

The drag of the water slows a boat down. 'Hydrofoils' solve this problem with special wings or foils attached to the hull by struts. The foils move underwater, working like plane wings to lift the boat. Because only the foils dip in the water, hydrofoils can travel very fast, often at speeds of more than 90 km/h.

Surface-piercing hydrofoil

Fully submerged hydrofoil

● Sail power

Sailing ships rely on wind power to drive them along and can sail in almost any direction except directly into the wind – because the wind does not push the sails but sucks them. As the wind blows over the curve of the sail, it speeds up and its pressure drops, creating suction in the same way that an airplane's wings create lift *(see pg20 [n7])*. However, the sail must be kept at exactly the right angle. Sailors let the sail swing round until the angle is right then hold it taut (tight) with ropes.

Read further › air pressure
pg20 [p2]; pg21 [b31]

A drag boat can cover 0.4 km in under 5 seconds

In the air

AIRPLANES ARE the fastest means of transportion, able to make journeys in just a few hours that would take days by road or sea. Most modern airplanes are powered by jet engines that can propel some military craft along at speeds of more than 3000 km/h, three times the speed of sound. Helicopters rely on their rotor blades to hover in the air. Not all air transport needs an engine – hot-air balloons use hot air and gas to stay up.

Swashplate controls angle of blades

Jet turbine engine

Rotor blade

To fly up or down, pilot alters angle or 'pitch' of main rotor blades with 'collective pitch' control. When blades cut through the air almost flat, they give no lift and helicopter sinks. To climb, pilot steepens pitch to increase lift

To fly forwards or backwards, pilot uses 'cyclic pitch' control to vary rotor pitch as blades go around from one side to the other

▲ The Sikorsky UH-60 Blackhawk can carry up to 12 soldiers and has two crew members.

● On the wing

Lift

Drag

Thrust

Weight

Flap

Rib

Spar

Trailing edge

Aileron

Tip

Leading edge

An aircraft's wings are lifted by the air flowing above and beneath them as they slice through the air. Because the top of the wing is curved, air pushed over the top is forced to speed up and stretch out, reducing its pressure and pulling from above. The curved shape is called an aerofoil. Under the wing, the air slows down and bunches up, and pressure here rises, pushing the wing up. How much 'lift' this creates depends on the angle and shape of the wing, and how fast it moves through the air.

▲ Thrust and lift are the main forces that enable aircraft to fly.

▶▶ Read further › lifting
pg15 (b30); pg19 (p23)

● Taking to the skies

Helicopters are able to take off vertically or hover in the air for long periods. They owe their versatile flying ability to big rotor blades, which are like long, thin wings that whirl around at high speed, slicing through the air and providing lift. The blades also act like huge propellers (see pg18 [n4]), hauling the helicopter upwards or backwards.

▶▶ Read further › propellers
pg18 (i13)

IT'S A FACT

• A jumbo jet, sucks in one ton of air every second.

• In December 1986, the American experimental plane, Voyager, flew right round the world without landing once.

● Check it out!

• http://travel.howstuffworks. com/airplane.htm
• http://travel.howstuffworks. com/airplane3.htm

Tail rotor drive shaft

FASTEST JETS

Jet, year flown	Speed
SR-71A Blackbird, 1976	3529 km/h
Mikoyan E-66, 1959	2387 km/h
F-100 Super Sabre, 1955	1323 km/h
F-86 Sabre, 1948	1080 km/h
Gloster Meteor, 1946	990 km/h

▶▶ **Read further › thrust**
pg20 (p2)

● How jets work

The simplest jets, called turbojets, work by pushing a jet of hot air out of the back to thrust the plane forward. Engines like these were used on the supersonic (faster than sound) airliner called Concorde, and some very fast military jets *(see pg21 [r28])*. Most airliners use quieter, cheaper turbofans that combine the hot-air jet with the draft from a whirling, multi-bladed fan to give extra thrust at low speeds.

● Hot-air balloon

Hot-air balloons get their lift by filling a huge bag of very light material with hot air from a gas burner. Because hot air is lighter and less dense than cooler air, the bag simply floats up, carrying the basket and passengers with it. As the air cools, the balloon begins to sink. To maintain height, the balloonist relights the burner to warm up the air in the bag again.

▲ *To descend quickly, the balloonist pulls a cord to let warm air escape through a vent in the top of the balloon.*

▶▶ **Read further › air pressure**
pg19 (b29)

Main shaft

Gases roar past exhaust turbines

'Bypass' air from main fan flows around engine core

Hot gases rush out of engine

Main fan sucks air in

Air is squashed by compressor turbines

Jet fuel is sprayed on to air, and continuous explosion happens

Afterburner burns leftover fuel in exhaust gases

▲ *In a typical turbojet, exhaust gases roar out of the back of the engine at more than 1600 km/h.*

Fin (tail)

Rudder turns plane left or right

Jet exhaust

Whole tailplane tilts as elevator to angle plane up or down

Single jet engine within fuselage (main body of plane)

Fuel tanks in wing

Air intake for jet engine

Cockpit

Radar in nose cone to detect other aircraft in the air

▲ *The jet-powered American F-16 Fighting Falcon is used by more than 14 countries worldwide.*

● Soaring forward

Today, nearly all military planes are powered by jet engines. The huge thrust of jet engines means wings can be smaller than on propeller-driven planes. They are swept back to allow the planes to fly fast with minimum air-resistance. The plane's path is directed by control surfaces – movable panels on the wings, tailplane and fin.

▶▶ **Read further › air-resistance**
pg19 (m32)

The Hughes H4 **Spruce Goose** flying boat of 1947 had the largest wingspan of any plane – 97.51 m

Space walk

TRAVELLING INTO space poses many problems. First, extremely powerful rockets are needed to accelerate a spacecraft up into space and free it from Earth's gravity. Second, out in space, spacecraft must survive in an environment unlike any conditions on Earth. Third, distances in space are vast, so the spacecraft must take everything it needs for a very long journey. However, stations in space can help the craft along the way, and astronauts are given special equipment to help them survive without oxygen and in extreme temperatures.

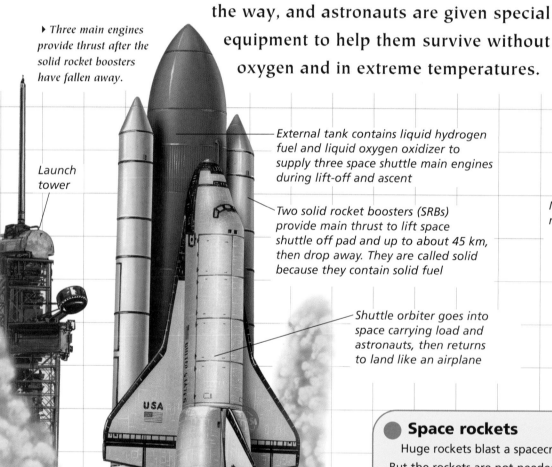

▶ *Three main engines provide thrust after the solid rocket boosters have fallen away.*

Launch tower

External tank contains liquid hydrogen fuel and liquid oxygen oxidizer to supply three space shuttle main engines during lift-off and ascent

Two solid rocket boosters (SRBs) provide main thrust to lift space shuttle off pad and up to about 45 km, then drop away. They are called solid because they contain solid fuel

Shuttle orbiter goes into space carrying load and astronauts, then returns to land like an airplane

Main living module

Solar panels

Main truss or 'skeleton'

Orbiter's three main engines fire at launch

● Space rockets

Huge rockets blast a spacecraft up and away from the Earth. But the rockets are not needed once the spacecraft is travelling fast enough to escape Earth's gravity, so they fall away in stages. At launch, the shuttle towers 56 m in the air and weighs 2000 tons. Most of this falls away soon after launch to leave the small orbiter, which weighs less than 45 tons and is less than 24 m long, to continue into space.

▶▶ **Read further › fuel tanks**
pg17 (b31); pg19 (k22)

The first four space shuttle orbiters were named after old sailing ships – **Columbia, Challenger, Discovery** and **Atlantis**

1   2   3   4   5   6   7   8   9   10   11   12   13   14   15   16   17   18   19

International Space Station

The International Space Station is assembled by about 15 different nations, using parts taken up into space by space orbiters and other rockets. Its overall length is about the same as a soccer field (about 92 m). The station consists of huge banks of solar panels each the size of a tennis court (about 24 m) that convert sunlight into electricity to be used onboard. The station's total weight is more than 450 tons – slightly more than a fully-loaded jumbo jet. Construction had began in 1998.

▶▶ **Read further ›space satellites**
pg27 (k22)

WORK IT OUT

• Future spacecraft may use sails to catch the solar wind – the stream of high-speed radiation from the Sun.

• It would take nine months to reach our nearest planet, Mars.

▶▶ **Read further › seeing space**
pg29 (b22)

Space wear

Outer space has very different conditions to those on Earth, and astronauts have to wear specially-designed spacesuits in order to survive in space. Spacesuits not only provide astronauts with vital oxygen for breathing, but also shield them against temperatures that vary from 120°C in the Sun to –100°C in the shade. Spacesuits have many layers to protect against cosmic rays and very fast particles of space dust called micro-meteoroids. The suits also create pressure to stop the astronaut's blood and body fluids boiling in the vacuum of space.

Ports for spacecraft docking

Detachable science modules and laboratories

Oxygen supply through an umbilical cord from spacecraft

Water pipes inside suit maintain body temperature

Backpack containing lithium hydroxide canisters to soak up astronaut's waste breath

Middle layer of suit inflated to provide pressure to stop body fluids from boiling

Multiple outer layers provide protection from micro-meteoroids

Light poly-carbonate helmet with anti-glare visor

Special tapered joints allow astronaut to bend hands, arms, knees and ankles

Inner layers of fabric keep astronaut warm

Reflective coatings provide protection from radiation

Check it out!
• http://spaceflight.nasa.gov/station/

A single spacesuit costs about $11 million to make!

a b c d e f g h i j k l m n o p q r s t u v w

Electronic brains

COMPUTERS HAVE entered our lives both at home and at work and can be used for anything from helping with homework to landing a spacecraft *(see pg22 [p6])*. Over the years, there has been a vast increase in the amount of data computers can store – and the speed with which they handle it. The advances in the computer world have even led to virtual reality – a way of creating artificial situations – being used for entertainment or business.

Read further > undersea wrecks / data
pg19 (b29); pg25 (b22)

WORK IT OUT

• In May 1997, IBM's Deep Blue computer beat the reigning World Chess Champion, Garry Kasparov.

• A 'byte' consists of eight on/off instructions or bits in a computer: 1024 bytes make 1 kilobyte (KB); 1024 kilobytes make 1 megabyte (MB); 1024 megabytes make 1 gigabyte (GB); 1 trillion bytes make 1 terabyte (TB).

Special eyepieces show slightly different views of an image to each eye to create illusion of real space

Sound effects are played in stereo through earphones, enhancing experience to make it seem even more realistic

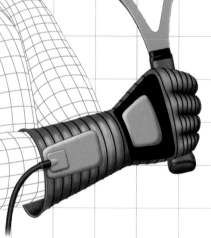

Is that for real?

Computers can sometimes fool our senses into thinking something is real using virtual reality (VR) systems. VR sends data (information) to our senses that closely mimic real scenes. A computer or robotic vehicle *(see pg19 [i32])* is sent to explore a situation such as an undersea wreck, or is programmed to show a fictional situation such as a tennis match. We can see what the computer sees by wearing a special headset.

IT'S A FACT

• The world's most powerful computer is the Earth Simulator, built at Kanazawa in Japan to help predict earthquakes.

• IBM are planning to build the world's biggest computer, called Blue Gene, to research proteins.

Check it out!

• http://www.computer.howstuffworks.com/computer-memory.htm

• http://computer.howstuffworks.com/bytes.htm

▲ *A number of pressure and flex (bending) sensors in the glove pick up wrist, hand and finger movements and send these along the cable to the computer. The computer analyses these movements to determine if the player had moved in the correct way to hit the 'virtual' ball, which can only be seen through the eyepieces.*

The speed and power of computers doubles, on average, every 18 months

Hardware and software

The actual material from which computers are made is called hardware. Equipment such as the keyboard, screen, mouse, printer, scanner and CD-writer are called peripherals – they sit outside the casing. The instructions or program that make the computer work is called the software. Software packages range from simple text and scanning images to the more challenging creation of graphics and special effects.

10100100 1010

100010100101

010101010010

Read further > storing data
pg33 (m30)

Zeros and ones

Since electronic circuits can only be switched on or off, computers store and handle data using a 'binary' system. This system turns all data into a code of 0s and 1s, or offs and ons. For example, the number five is 0101 in binary code, or off-on-off-on. Each 0 or 1 is called a binary digit or 'bit,' and they are grouped together in 'bytes.' Electronic circuits store and process data in this way.

101
01001 0010
0101001010010

Read further > electrical signals
pg26 (l2); pg27 (b22)

Plastic casing protects components

Chip

◄ *The main part of a computer is the central processing unit, containing a microchip that is no bigger than a fingernail.*

Wire 'feet' are connections that attach to other parts in computer

Microchips on circuit board

Main computer casing, to protect all components inside

'Read-only memory' (ROM) is computer's basic working instructions installed in computer at factory

Flat screen monitor

Digital camera or Webcam to record images and send messages or photos over Internet

Inside central processing unit (CPU) is a powerful microchip – a tiny block of complex electronic circuits – that is 'brain' of computer and carries out main tasks

Random-access memory (RAM) temporarily stores data as it is being used

Data can be stored as laser guided pits on a CD or DVD that can be inserted into computer in CD or DVD drive (Reader)

Roller sensors

Moving mouse rolls the ball that turns sensors so on-screen cursor moves

Keyboard to type on or control computer functions

The Internet was developed in the 1960s by the US army to link their computers

Instant contact

TELECOMMUNICATIONS MAKE it possible to see and talk to people almost anywhere in the world. Some communications such as phone calls, faxes or e-mail can be one-to-one. Radio and television (*see pg29 [f33]*) programs are broadcast to millions of people. Cable TV and web-casting (Internet broadcasting) combine both elements. Yet all these forms of communication work in much the same way.

IT'S A FACT

• The Internet grew during 2002 at the rate of 300 new pages or screens of information being added every minute.

• Many cell phones include a digital camera for sending live video pictures.

Phones on the move

Mobile or cell phones use low-power radio waves to send messages. Areas across the world are divided into many small sectors called cells, each with an antenna (*see pg27 [s24]*) that picks up signals from phones and sends them out. Because there are so many antennae spread across the world, millions of people can use cell phones at once.

Small loudspeaker in earpiece

Mode or function keys

Small microphone in mouthpiece

Numerical keys

Land lines

Telephones convert sound into an electrical signal. When you speak into a phone, the vibrations of your voice move a tiny microphone that alters an electrical current in strength. This creates an electrical signal, which is sent to the receiving phone. In the receiver the varying signal works a loudspeaker, which vibrates the air and recreates the sound of your voice. Today, many signals are sent as pulses of laser light along special glassy fibers called optical fibers. Signals are also sent as radio waves or microwaves through the air – bouncing off satellites (*see pg26 [p18]*) in space.

Read further > signals
pg15 (b22)

Many communications, such as mobile phone calls and e-mails, are sent on or relayed by satellites in space

Signals from individual transmitters are sent on from a telephone exchange or a service provider

Computer data is translated by a modem into signals that can be carried along phone lines

Read further > electromagnet
pg15 (b30)

TV and radio signals are either broadcast as pulses of radio waves, sent direct via cables or broadcast from satellites

The telephone network is the world's biggest machine

1   2   3   4   5   6   7   8   9   10   11   12   13   14   15   16   17   18   19

Instant letter

Fax is short for 'facsimile,' which means 'copy.' A fax machine scans a document under a light past a line of sensors. The white parts of the document reflect light so switch on the sensors; the dark parts do not reflect light so switch off the sensors. Thus, a pattern of ons and offs is created and sent to the receiving fax. The receiving fax machine uses heat sensors that pick up the pattern of electrical signals coming through and creates the same pattern on the heat-sensitive paper. More modern machines print on plain paper, using charges of static electricity to attract the toner powder to the paper.

Telephone

Dial pad for typing in fax number

Phone socket that enables fax to connect to a phone line and send and receive messages

Image received is printed on paper from internal roll

Image to be sent is fed through a scanner

Drive rollers, to run document under scanner

Scanner sensors

Read further › on / off patterns
pg25 (b32)

Bouncing signals

Any telecommunications message needs a transmitter such as a phone; a communications link such as an antenna or satellite; and a receiver or destination, such as an e-mail address or receiving telephone number. The message can travel through electrical or optical (fiber optic) cables, as microwaves or as radio waves, to arrive at its destination.

Read further › telephones
pg26 (k2)

Communications travel via satellites and are beamed up and down from antenna dishes on ground

Telephones link into phone network by a direct cable link. Mobile phones link through the air to local relay towers by radio waves

WORK IT OUT

• Billions of cell phones are in use around the world. Average use for a cell phone in the US is 1 1/2 years – the US Geological Survey estimates less than 1% of these will be recycled.

• Most of the 5000 satellites circling the Earth in space are used for telecommunications. Communications satellites use a special orbit, called a geo-stationary orbit, which keeps them in the same place above the Earth constantly.

Electronic letters

A fast and convenient form of contact today is electronic mail or e-mail. The message is typed up on a computer or even on some mobile phones and is sent to another e-mail address along a phone line, via a modem, to the central computer of the ISP (Internet Service Provider). The message is stored here until the recipient connects or 'logs in' to a computer to receive it.

Read further › computers
pg25 (b22)

In 2004, 500 billion text messages were sent.

Super vision

MICROSCOPES THAT magnify objects too tiny for our eyes to see and telescopes that magnify things that are far away have been used since the 1600s. Photography has been recording accurate pictures for more than 160 years. Today, microscopes can show fantastic pictures of minute micro-organisms or even atoms and telescopes can reveal the most distant galaxies in the universe. Cameras take better quality photographs than ever, and digital technology enables pictures to be scanned into a computer to be enhanced and even sent across the world instantly via e-mail *(see pg26 [s5])*.

Eyepiece lens

Pentaprism turns image right way round for eyepiece

Zoom ring

Objective lenses gather light rays of scene together and make them converge so that image is smaller

Aperture (hole for incoming light)

Mirror reflects light up to pentaprism

Light from scene

Lenses

Focus ring

▲ Single-lens reflex (SLR) cameras and simpler compact cameras use a different technology than more recent digital cameras, which record images electronically on a microchip.

Snap happy

The lens inside a camera is a disk of glass or plastic, specially shaped to focus (bring together) all the light from the scene and create a tiny image (picture) inside the camera. In most cameras, the image is recorded on film coated in chemicals that react to the light that comes in through a hole (aperture). The mirror reflects the light up to a prism and then to the eyepiece for the photographer to see the image. To take the picture, the mirror swings out of the way so that light falls on to the photographic film behind it for a split second.

WORK IT OUT

• A microscope's magnification is limited by the smallest wavelength of light, which is about 4000 angstroms (a millionth of a centimeter). The wavelength of electrons used in electron microscopes is just 0.5 angstroms.

• Individual atoms can be seen with the scanning tunnelling microscope (STM), which scans the surface of materials by tunnelling an electric charge into them.

▶▶ Read further > pixels
pg32 [q2]

The new Large Binocular Telescope at Mt Graham in Arizona has the world's largest telescope mirrors, 8.4 m across

1 2 3 4 5 6 7 8 9 10 11 12 13 14 15 16 17 18 19

a
b
c
d
e
f
g
h
i
j
k
l
m
n
o
p
q
r
s
t
u
v
w

Farsighted

Telescopes help you see distant things better by focusing (concentrating) the light onto a small area, then magnifying the focused image so you can see it. In 'refracting' telescopes, a large lens focuses the image by bending or refracting the rays, then a smaller lens, called the eyepiece, magnifies it. In 'reflecting' or mirror telescopes, the image is focused by a large dish-shaped mirror onto a smaller mirror and through the eyepiece. Astronomers now use amazingly powerful telescopes to see galaxies billions of light-years away. The image from these galaxies is picked up by light-sensitive pixels and fed into a computer, where it can be electronically enhanced.

Read further › going into space
pg22 (r11); pg23 (b22)

◀ *Telescopes allow astronomers to see stars not visible to the naked eye.*

▾ *Reflecting telescopes have a curved main mirror.*

Eyepiece lens

Incoming light

Secondary mirror

Flat mirror

Incoming light

Focused light

Reflected light

Objective (front) lens

Main mirror

Eyepiece lens

◀ *Refracting telescopes use only lenses, not mirrors.*

Seeing microbes

Powerful 'electron' microscopes work differently from optical microscopes. Instead of using lenses to magnify light, they fire streams of particles – electrons – at the specimen (the object being studied) and show the result on a monitor. In a transmission electron microscope (TEM), the specimen is so thinly-sliced – usually less than one-hundredth of a millimeter – that electrons can pass through to give a shadow, which is picked up by detectors under the image. Chemicals are added to partly block the electron beam and make the shadows stand out more strongly. TEMs can magnify things up to 1 million times.

▸ *A scanning electron microscope (SEM) scans across the surface of the bug to get a very detailed, almost three-dimensional surface view.*

Read further › television
pg33 (b22)

Round lenses

Glass lenses are shaped to bend light rays in a particular way. Sometimes they are made dish-shaped or 'concave' – thin in the middle and thick round the rim. Light rays passing through a concave lens are bent outwards, so spread. This means when you see something through a concave lens it looks smaller than it really is. Other lenses bulge outwards or are 'convex' – fat in the middle and thin at the rim. Light rays passing through a convex lens bend inwards, so come together or converge. This means that things look bigger when viewed through a convex lens. The point where the converging light rays meet is called the focus.

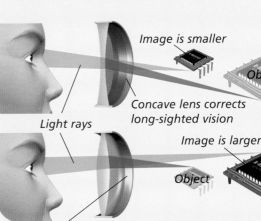

Image is smaller

Object

Concave lens corrects long-sighted vision

Light rays

Image is larger

Object

Convex lens corrects short-sighted vision

Read further › light rays
pg30 (d15)

Some microscopes can show individual atoms (tiny particles) of substances such as gold

Lasers and X-rays

JUST OVER a century ago, Röntgen discovered how to use X-rays to scan inside the human body. Since then, many kinds of scanners have been developed using lasers – powerful beams of light – not just for use in medicine but in shops to 'read' bar codes that control the stock levels of merchandise. Lasers are so versatile that they are used in CD and DVD players and even to cut very strong materials by burning through them. Lasers are also used to create images, including three-dimensional images called holograms.

How lasers work

Laser light is made by feeding some form of energy, such as ordinary light or electricity, into a substance called the active medium. In a ruby red laser, the active medium is a rod of ruby crystal. Electric sparks cause the tiny particles inside the crystal to vibrate and give off tiny packets of photons of light. The light builds up and is bounced between mirrors at each end. The energy becomes so strong that it escapes from the end as an intense pulse of light.

Read further › colours
pg35 (c33)

Particles bounce to and fro in ruby crystal

Silvered mirror

◄ Waves of light emerge at one end of a laser to create a very bright light.

Casing

Excited atoms fire off 'photons' (little packets of light)

Part-silver mirror

When laser beam has sufficient power it bursts from one end of crystal

Rays bounce off mirrors at both ends, building up energy

Electric spark excites atoms in ruby

▼ Industrial 'gas' lasers can cut accurately through sheets of metal, while lower-powered ones cut up rolls of textiles for making clothes.

IT'S A FACT

• Laser light is even brighter than the light from the Sun.

• Laser beams are so straight that they can be aimed at a tiny mirror left by astronauts on the Moon (384,401 km away) to measure the Moon's distance from Earth.

Check it out!

• http://www.enter.net/~holo studio/holo/kids.html

• http://science.howstuffworks. com/laser.htm

Laser beams can carry 1000 times more information than microwaves

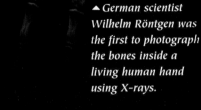

▲ *German scientist Wilhelm Röntgen was the first to photograph the bones inside a living human hand using X-rays.*

WORK IT OUT

• Magnetic Resonance Imaging (MRI) scans take pictures of the body using powerful magnets to line up protons (parts of atoms), then recording the many radio signals sent out by the protons as they realign when the magnets are turned off.

• Positron Emission Tomography (PET) scans show how fast a body part is working. The scans are carried out by putting slightly radioactive substances into the blood. These substances emit positrons (positively charged electrons) that send out tiny rays for the scanner to pick up.

● Seeing through people

X-rays – waves of radiation too short for the eye to see – enable pictures to be taken of the inside of the body. X-rays pass straight through the body, yet turn photographic film black, just like ordinary light. Tissues that block their path, such as bones, show up white on the film. However, the rays must be very carefully controlled – in large amounts they can be dangerous.

▶▶ **Read further › light and dark**
pg27 (b22); pg32 (q2)

● Solid pictures?

Holograms – three-dimensional images that change when viewed from different angles – are made by splitting a laser beam in two. One half – the reference beam – goes straight to the film. The other half bounces off the object and lands on the film, breaking up the rays of the reference beam. When the two beams are reunited, the film shows how the broken pattern differs from the neat pattern of the reference beam.

▶▶ **Read further › patterns**
pg25 (b22)

● Bar code reader

When you buy something from a store, it will probably have a band of black and white lines printed on it called a bar code. A barcode is an effective way of keeping a constant track of everything sold – and providing customers with instant pricing information when they reach the register. The pattern of lines is a code for the product's number listed in the memory of the store's computer. The code is in binary form – that is a code for a number in terms of 0s and 1s or 'ons' and 'offs.' When the bar code reader runs its laser beam over the code, it detects the code by shining a light and picking up reflections from the white stripes. The reader then sends an electrical signal to the computer, which instantly displays the product details.

Bar code reader

Laser bounces back to relay details

The laser quickly scans the lines

◀ *The laser beam inside a bar code reader bounces back and forth, sending information to the computerized cash register.*

▶▶ **Read further › language**
pg25 (b32)

Laser-light photographs can capture very high-speed action with exposures of just a few trillionths of a second

Sound and vision

TELEVISION BRINGS pictures and sound from all around the world right into our homes, letting us watch events such as top sports games as they happen live across the world. Not only can we watch such events on television, but we can also record many of our own events, using personal camcorders. These recordings can be saved on video tapes or disks to play back later. Other recordings such as music or films can be stored on CD or DVD to be played many times.

IT'S A FACT

• A TV camera converts a scene into electrical signals using three sets of tubes that react to red, green and blue light.

• CDs record data digitally – that is as a simple on/off code of bits – like computers and fax machines.

▼ *A camcorder records images on a small reel of magnetic tape (videotape) or in microchip memory circuits.*

Objective lenses

CCD (charge-coupled device) has an array of pixels that detect light

Electric motors and gears to move lenses for focusing (making the image sharp and clear) and zooming (bigger or smaller)

Eye piece

Eyepiece display screen

Monitor screen showing what is being viewed or recorded

Control buttons

Camcorders

Digital camcorders have a lens for projecting a picture that can be recorded using a tiny patch of light-sensitive cells, or pixels. Where the light in the picture is bright, the cell sends out a brief electrical pulse; where it is dark, the cell stays off. As the picture changes, this creates a pattern of electrical pulses. This pattern can then be stored in the camera's memory for playback later, or be sent instantly to the camcorder's viewer or a TV screen to be watched as it is viewed or recorded.

▶▶ **Read further › animated film** **pg35 (b25; k27)**

WORK IT OUT

• The spiral of bumps on a CD are less than half a millionth of a meter wide. The bumps are just 125-billionths of a meter deep.

• Because there is a limit to the resolution (sharpness) of conventional 'analog' TV, more and more broadcasters are switching to high definition digital pictures like those on computer screens, which can be up to 10 times sharper.

If the data on a single DVD were not digitally compressed, it would take a whole year to send it over a phone line

| 1 | 2 | 3 | 4 | 5 | 6 | 7 | 8 | 9 | 10 | 11 | 12 | 13 | 14 | 15 | 16 | 17 | 18 | 19 |

On the tube

Older televisions have a slightly curved screen, which is the front end of a device called the tube – like a giant light bulb. Inside the rear of the tube are 'guns' that create the picture by firing nonstop streams of electrically-charged particles, called electrons, at the back of the screen. Where they hit the screen, they make it glow by heating up its coating of phosphor dots. When we look at the screen, we see thousands of these glowing spots of phosphor in the form of a picture. The broadcast or recorded signal controls where the beams target the screen, making it glow and creating the picture.

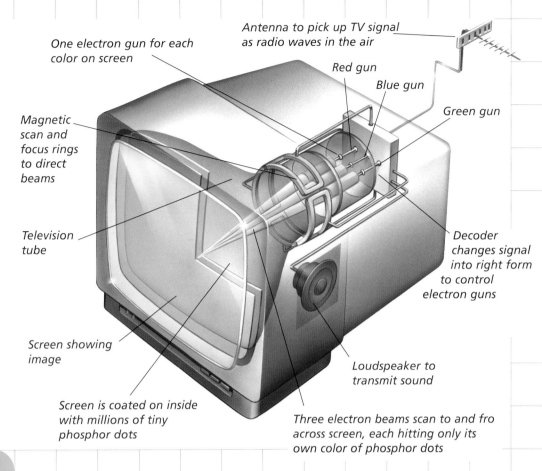

One electron gun for each color on screen

Antenna to pick up TV signal as radio waves in the air

Red gun

Blue gun

Green gun

Magnetic scan and focus rings to direct beams

Television tube

Decoder changes signal into right form to control electron guns

Screen showing image

Loudspeaker to transmit sound

Screen is coated on inside with millions of tiny phosphor dots

Three electron beams scan to and fro across screen, each hitting only its own color of phosphor dots

Read further › electrical particles
pg29 (k22)

DVDs

Digital Versatile Discs (DVDs), are an efficient way of storing all kinds of data, from music and films to video games. Made of plastic, coated with acrylic and sprayed with aluminium, DVDs store data as a spiral track of tiny pits pressed into the otherwise flat (land) surface. When the DVD is played, a laser beam 'reads' these pits by scanning the underside of the disc. As light hits land it is read as a binary 1. Light that hits a bump is read as a binary 0.

▾ *DVDs and CDs work in the same way but a CD holds about one-seventh of the data of a DVD.*

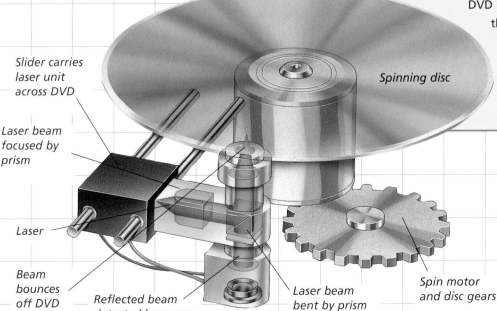

Slider carries laser unit across DVD

Spinning disc

Laser beam focused by prism

Laser

Beam bounces off DVD

Reflected beam detected by sensor

Laser beam bent by prism

Spin motor and disc gears

Check it out!

• http://www.howstuffworks. com/tv.htm
• http://electronics.howstuff works.com/camcorder.htm

One CD can contain about 600 megabytes of data – about 375,000 pages of ordinary text

Mass media

MANY KINDS of different media are used for communicating with large numbers of people, whether for entertainment, information or advertising. Newspapers, books, magazines, films, cartoons and the Internet are just some of the many media available to the public. The printing press has enabled written media to reach more people, faster than before. Computers now combine different media to create 'multimedia.' This combination can be used to create video games, CD-Roms, interactive TV and computer-animated films that are more realistic and ambitious than ever.

▲ *Animatronic models such as dinosaurs are usually filmed against a blue screen, while a background such as a wood is filmed separately. The blue screen is later replaced by the real background.*

● IT'S A FACT

• The lightsabers in the *Star Wars* movies are filmed using aluminium dummies, which are then replaced by light on a computer, frame by frame.

• Repeat print-runs of books can be made by simply inserting electronic 'smart cards' into the printing machine's computer.

● Straight off the press

Computer technology has revolutionized printing. In the past, pages were printed by wiping ink over metal type (words and letters), which had to be made from scratch using hot metal, or even assembled painstakingly from a store of letters. Now an exact picture of all the text and illustrations can be created on a computer screen. This computer picture is then used either to make a film that can set up the printing photographically, or directly control lasers or jets of ink as they print on to the page.

▶▶ Read further › laser light
pg30 (d15)

▼ *The full-color lithoprinter makes colored images out of four colors of ink – cyan (C), magenta (M), yellow (Y) and black (K) – as tiny dots on the page. Four-color or CMYK printing is used to produce most magazines, books and newspapers. The fifth press contains a special mix of color or 'text black' for printing words.*

Reel of paper can be changed while printing press is still running so no time is lost during printing

Inking rollers make sure ink is spread evenly on paper. Surfaces of roller can be made from rubber or metal

'Paper web' is a very long, continuous piece of paper that runs through press to be cut into separate sections at end of the press

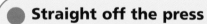

Transfer roller holds paper tightly in place to ensure everything is printed in the right place and that paper does not move between printing presses

Image that needs to be printed comes from a plate that is placed in press and transferred to rubber blanket roller, which rolls image over paper evenly

Drier's heated plates dry ink so that it does not smudge when paper is folded

Bringing dinosaurs to life

Dinosaurs died out 65 million years ago, but amazingly lifelike recreations in films, on TV and in museums enable us to see what they were really like. These are called 'animatronics' and are basically mechanical puppets. From walking to breathing, each movement is created by electric motors and hydraulics, controlled remotely by a team of operators. The initial designs are hand-drawn and then reproduced by computers, which in turn guide the machine to make a mold. This mold is used to make a body shell from stiffened foam rubber.

▶▶ Read further > hydraulics
pg13 (f32)

▶▶ Read further > printing
pg34 (k2)

A world in three colors

Almost every color can be printed by combining just three 'primary' colors of ink. Three 'separations' – versions of the picture in terms of its red, green and blue content – are created. Layers of yellow, magenta and cyan inks then soak up the right amounts of blue, green or red from the light falling on the page

▶ *Primary colors combine to make white.*

▶▶ Read further > camcorders
pg32 (q2)

Living pictures

Animation means creating the illusion of movement by showing a sequence of subtly-changing still pictures. In the past, pictures were drawn painstakingly by hand, which could take a long time. The use of computer-generated animations has sped up the process and enabled a startling amount of realism by scanning real movements into the camera to act as a base for the animation. Since the computer can store a complete three-dimensional image of all the characters and the way they move, the animator only has to drag the character into position on screen to make a new frame (scene).

▲ The model of a character such as a monster is photographed in one pose. A computer scans the photograph and plots the positions of each tiny part of the model's surface into its memory.

◀ The model is moved into another position and photographed again. The computer scans this new image, compares it with the first, and 'fills in' the in-between images so that they differ only slightly from each other.

Folding unit folds paper so it can be taken off press and packaged immediately

Folded pages are cut to right size and stapled or stitched to hold batches together for easy transportation out to retailers

● WORK IT OUT

• For the movie *Jurassic Park III*, a life-sized animatronic of Spinosaurus – one of the largest meat-eating dinosaurs that ever lived – was made measuring 13.3 m long and weighing more than 12 tons.

• The first movie to be made entirely with computer-generated animation was Disney's *Toy Story* in 1995.

To achieve extra realism, animated movies such as **Shrek** base movements on real animals and people

Glossary

Abutment A solid structure that supports the end of a bridge or takes the load of an arch.

Aerofoil A curved surface such as an aircraft wing, which provides control in the air.

Animation The making of films by photographing a sequence of drawings or models so that they appear to be moving.

Antenna A metal rod or wire for sending and receiving radiowaves or microwaves.

Ballast Heavy material carried in the bottom of a boat or balloon to weigh it down and give it extra stability.

Bascule A bridge with a roadway that can be lifted to allow tall ships to pass through.

Binary Consisting of two parts or numbers. A computer code of offs or ons, or 0s and 1s.

Buoyancy How well something floats.

Cantilever A bracket that supports a shelf or balcony, or a bridge with rigid arms projecting out from piers to meet in the middle.

Compression Squeezing, especially of air or gas, inside a cylinder.

Computer Electronic device that processes, stores and represents data in a new form according to instructions called a program.

Concave Curved inwards like a dish.

Convex Curved outwards like a ball.

Crankshaft A shaft driven around by an arm called a crank, at right angles. In a car engine the crankshaft is turned by the moving rods attached to each piston.

Cylinder In a car engine, a small tube in which the piston goes up and down.

Drag The way in which air or water slows down an object moving through it. This is sometimes called air or water resistance.

Electromagnet Powerful magnet that works when an electric current is passing through it.

Electron The smallest of the particles of an atom. Electrons circle round the atom's nucleus and have a negative electrical charge. Electricity is the movement of electrons.

Foundations The solid base of a building, often below ground to support the structure.

Fulcrum The point at which a lever turns.

Fuselage The tube-shaped body of an aeroplane, excluding the wings.

Gear Wheel that turns another wheel at a different speed or force.

Girder Large strong beam, usually made of steel, to support a building or bridge.

Gravity The force of attraction between every bit of matter in the universe. Gravity makes things fall, holds things to the ground, and keeps the Earth and other planets circling round the Sun.

Hologram Special 3D photograph created by splitting a beam of laser light into two.

Hydrofoil A slightly curved flat surface like a ski that lifts a boat out of the water and allows it to glide along on top of the water.

Inclined plane A slope that makes lifting easier by sliding or rolling things.

Induction The induction stroke in a car engine is the movement of the piston that draws new fuel into the cylinder.

Knot A unit for measuring speed at sea or in the air. A knot is about 1.85 km/h.

Lift Force to lift an aircraft or hot-air balloon.

Maglev Train without wheels that glides above the ground, lifted by magnets.

Momentum Makes an object move at the same speed and in the same direction.

Monorail Railway line with only one rail, often running overhead.

Optical fibre Special glass cable that transmits signals using laser light.

Pier Solid support pillar for a bridge or platform built out into water.

Pile Column driven into the ground to support a building or bridge.

Positron A particle identical to an electron but with a positive electrical charge.

Prism Wedge of glass that splits white light passing through it into different colours.

Pulley Wheel with a grooved rim that guides a rope to make it easier to lift things.

Rack and pinion A range of gears in which a toothed wheel (the pinion) runs along a straight row of teeth (the rack).

Radiation Energy emitted from particles as electromagnetic waves such as light and radio waves, or as radioactive particles.

Shield Special drum-shaped machine for boring out tunnels.

Submarine A large boat built for travelling under the sea for long periods.

Submersible A small craft built to travel short distances under the sea or deep down, carrying just one or two people or entirely robot-operated.

Telecommunications Sending and receiving of sound, pictures or data over long distances, by radio waves or by electrical or optical cables.

Thrust The force driving something along.

Virtual reality A computer recreates a three-dimensional space, so that it seems as if you are actually in the space.

X-ray High energy electromagnetic radiation used to take pictures through the body of bones and other structures.

Index

Entries in bold refer to main subjects; entries in italics refer to illustrations.

The publishers would like to thank the
following artists who have contributed to this book:
Kuo Kang Chen, James Evans, Mark Franklin, Alan Hancocks, Robert Holder,
Rob Jakeway, Janos Marffy, Helen Parsley, Terry Riley, Martin Sanders,
Peter Sarson, Rudi Vizi, Paul Williams

The publishers wish to thank the following sources for the photographs used in this book:
p13 (t/r) Vince Streano/Corbis, p17 (t/l) Patrick Ward/Corbis, p26 (c/r)
Nokia, p31 (b/r) TESCO, p34/35 (c/t) Universal

All other photographs are from:
Castrol, Corel, DigitalSTOCK, PhotoDisc, MKP Archives

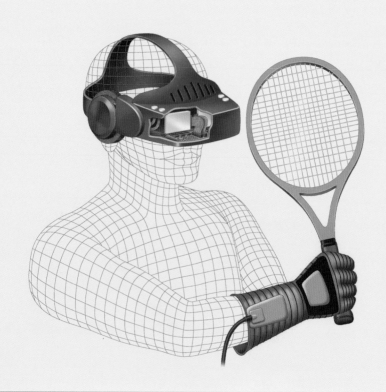